José Martin Glavas

Physical and Chemical Agents Analysis in Occupational Health

José Martin Glavas

Physical and Chemical Agents Analysis in Occupational Health

Air Quality, Ambient Noise and Thermal Load

ScienciaScripts

Imprint

Any brand names and product names mentioned in this book are subject to trademark, brand or patent protection and are trademarks or registered trademarks of their respective holders. The use of brand names, product names, common names, trade names, product descriptions etc. even without a particular marking in this work is in no way to be construed to mean that such names may be regarded as unrestricted in respect of trademark and brand protection legislation and could thus be used by anyone.

Cover image: www.ingimage.com

This book is a translation from the original published under ISBN 978-613-9-44010-8.

Publisher:
Sciencia Scripts
is a trademark of
Dodo Books Indian Ocean Ltd. and OmniScriptum S.R.L publishing group

120 High Road, East Finchley, London, N2 9ED, United Kingdom
Str. Armeneasca 28/1, office 1, Chisinau MD-2012, Republic of Moldova, Europe
Printed at: see last page
ISBN: 978-620-3-35814-8

Contents

A- FINAL WORK EVALUATION OF AMBIENT AIR QUALITY (MEASUREMENTS: PARTICULATE MATTER, MULTIGAS AND ENVIRONMENTAL QUALITY).

Ing. Qco. Jose Martrn Glavas

Campus Facultad de Ciencias Agrarias de la Universidad Nacional del Nordeste (UNNE) - City of Corrientes - June 2022

josemartinglavas18@gmail.com

SUMMARY

A technical report of air quality evaluation at the Campus of the Faculty of Agricultural Sciences of the National University of the Northeast (UNNE) to determine the quantities and concentrations of particulate and multigas pollutants, as well as environmental quality can help to determine if there is an environmental problem (pollution). Locating the source of these various pollutants can help determine effective methods of reducing them and improving air quality. These measurements are detailed below:

a)   Particulate Matter (PM) by means of a Particle Counter, Tenma, Model ST - 9880, which is used to measure the number and size of particles in the air. The data obtained is used to measure the contamination due to fine dust present on the campus. To calculate the data it sucks in air for a selected time and calculates the number and size of particles contained in the air. In doing so it considers equally the particle sizes 0.3, 0.5, 1.0, 2.5, 5.0 and 10.0 urn. The ambient contamination for a particle of the size selected by the user is indicated by a colour scale (see ISO Standard No. 14644 - 1: 2015 page 6); all measurements are within the green (permitted) colour range of this standard. In addition to the number of counted particles, the temperature, the humidity of the air, as well as the calculated dew point and the wet bulb temperature are indicated.

b)   Multigas through gas detector, Drager, Model X - am 2500 for continuous monitoring of the concentration of various gases in the ambient air, measured in ppm (parts per million). Independent measurement of up to 6 gases corresponding to the Drager sensors installed. They are: Oxygen (O_2), Carbon Monoxide (CO), Hydrogen Sulphide (SH_2), Methane (CH_4), Nitrogen Dioxides (SO_2) and Sulphur Dioxides (SO_2). The values found are within the Accepted Values of Annex III of the Regulatory Decree N° 351/1979 of Law N° 19587/1972.

c)   Environmental quality by means of the Mini Environmental Quality Meter, Sper Scientific, Model No. 850027; whose factors are measured through: Air Speed (meters/seconds), Air Flow (CMM), Fresh Wind (°C), Relative Humidity (%), Flash Point Temperature (°C), Wet Bulb Temperature (°C), Heat Index (°C), Barometric Pressure (hPa) and Altitude (meters). The Heat Index complies with the Heat Index Table of the U.S. National Meteorological Department (see Table No. 2 page 15).

Key words: Assessment, Quality, Air, Environmental, Industrial.

1. INTRODUCTION.

Air pollution is one of the most important environmental problems, and is largely the result of human activities. The causes of air pollution are diverse, but the highest rate is caused by industrial, domestic, agricultural, livestock and vehicular activities, among others.

Combustion, used for heat, power generation or motion, is the most significant pollutant emitting process. Other activities, such as smelting and the production of chemical substances, can cause deterioration of air quality if left unchecked.

Pure air is a gaseous mixture composed of 78 % N_2 (Nitrogen), 21 % O_2 (Oxygen) and 1 % of different total compounds such as Argon (Argon), CO_2 (Carbon Dioxide) and O_3 (Ozone). Air pollution means any change in the balance of these components, which alters the physical and chemical properties of the air.

The MPC (Time Weighted Maximum Allowable Concentration) or TLV (Threshold Limit Value) values refer to concentrations of substances which are suspended in air (see Annex III of Regulatory Decree No. 351/1979 of Law No. 19587/1972).

However, given the great variability in individual susceptibility, it is possible that a small percentage of workers may experience discomfort at some substances or concentrations at or below the Threshold Limit, while a smaller percentage may be more seriously affected by the aggravation of a pre-existing condition or by the onset of an occupational disease.

The CMP values are based on available information obtained through experience in industry, human and animal experimentation, and where possible, by a combination of all three. The basis on which CMP values are established may differ from substance to substance, for some substances, protection against health impairment may be a factor, while for others the reasonable absence of irritation, narcosis, annoyance or other forms of discomfort may be the basis for setting the CMP value. The health impairments considered refer to those that decrease life expectancy, compromise biological function, impair the ability to defend against other toxic substances or disease processes, or adversely affect reproductive function or developmental processes.

Definitions:

a. MPC (Time Weighted Maximum Allowable Concentration): Time-weighted average concentration for an 8-hour/day workday and a 40-hour workweek, to which it is believed almost all workers can be exposed repeatedly day after day, without adverse effects.

b. CMP - CPT (Maximum Permissible Concentration for Short Time Periods): Concentration at which it is believed that workers can be exposed continuously for a short period of time without suffering: 1) irritation, 2) chronic or irreversible tissue damage, or 3) narcosis to a degree sufficient to increase the likelihood of accidental injury, make it difficult to extricate oneself from a hazardous situation or to reduce work efficiency, and provided that the daily MPC is not exceeded. It is not a separate exposure emission limit, but rather complements the time-weighted average emission limit when acute effects of a substance whose toxic effects are primarily of a chronic nature are accepted. Short-term maximum concentrations are recommended only when toxic effects in humans or animals have been reported as a result of short-term, intense exposures.

The CMP - CPT is defined as the 15-minute time-weighted average exposure, which must not be exceeded at any time during the working day, even if the eight-hour time-weighted average is less than this Emit value. Exposures above the CMP - CPT up to the short duration exposure Emit value must not exceed 15 minutes in duration and must not be repeated more than four times a day. There must be at least 60 minutes between successive exposures in this range. An average exposure period other than 15 minutes may be recommended when justified by the observed biological effects.

c. MPC-C (Maximum Permissible Concentration - Ceiling Value (C): is the concentration which must not be exceeded at any time during an occupational exposure. In conventional industrial hygiene practice, if an instantaneous measurement is not possible, the MPC-C can be set for short exposures by sampling for a time not exceeding 15 minutes, except for those substances which may cause immediate irritation.

Particulate matter (PM) is a mixture of air-disturbing particles in suspension, of different physical and chemical characteristics and originating from a variety of sources and emission sources. Particles may act as a medium in which certain chemical reactions occur, condensation nuclei or elements capable of scattering, absorbing and emitting radiation.

Particulate matter is classified according to its diameter:

> PM 10 particles (smaller than 10 pm in diameter): These are mainly primary particles emitted directly into the atmosphere by natural phenomena and human activities; due to their size they tend to be deposited close to their place of origin and have a greater capacity to access the respiratory tract and therefore have a greater effect on the respiratory tract.

> PM 2.5 particles (less than 2.5 pm in diameter): These are mainly composed of secondary particles resulting from chemical processes or reactions in the liquid phase of substances such as NO2 (Nitrogen Dioxide), SO2 (Sulphur Dioxide), VOC (Volatile Organic Vapours), NH3 (Ammonia), etc.Due to their small size, they remain in suspension for a long time, travel long distances and are deposited in the deepest part of the respiratory system, becoming trapped and can generate more severe health effects.

<u>I alarm units for particle concentration</u>:

ISO Standard No. 14644 - 1: 2015 (see Table 1) refers to the classification of air cleanliness, according to this standard, exclusively in terms of the concentration of suspended particles. Also, for the classification according to this standard, particles with defined dimensions in the range of 0.1 to 5 pm (ppm = parts per million) are considered.

Counts of leftover particles, grouped by colour, green (permitted), yellow (caution) and red (danger), are displayed for each channel:

Table N° 1

Channel	Green (Permitted)	Yellow (Caution)	Red (Danger)
0.3 pm	0 ~ 100000	100001~250000	250001~500000
0.5 pm	0 ~35200	35201~87500	87501~175000
1.0 pm	0 ~ 8320	8321~20800	20801 ~41600
2.5 pm	0 ~ 545	546 ~1362	1363 ~2724
5.0 pm	0 ~ 193	194 ~483	484 ~966
10.0 pm	*0 ~ 68*	*69 ~ 170*	*170 ~ 3410*

<u>Properties of dangerous gases and vapours:</u>

Flammable and toxic gases, vapours and vapours can occur in many places. To deal with the toxic risk and the danger of explosion, gas detection systems are used.

Practically all gases and vapours are always dangerous. If gases do not exist in their familiar, breathable atmospheric composition, safe breathing may already be affected. Any gas is potentially dangerous, whether it is liquefied, compressed or in its normal state, the important thing is to know its concentration.

Basically there are three categories of risk:

- Ex - Explosion hazard due to flammable gases.

- Ox - O2 (Oxygen).

Risk of asphyxiation due to oxygen displacement.

Risk of increased flammability due to oxygen enrichment.

- Tox - Risk of poisoning by toxic gases.

Measurements of pollutants in the workplace are the only effective measure to determine whether personnel are exposed to chemical substances and/or particulate matter, which could cause deterioration of the health of exposed workers, leading to occupational diseases over time.

It is recommended to carry out such checks on a regular basis, as the working conditions may change over time due to various factors, such as: changes in production quantities, changes in the types of substances used to manufacture a product, changes in the ventilation systems of a plant, building modifications, etc.

2. DEVELOPMENT.

2.1- MEASUREMENT OF PARTICULATE MATTER IN AMBIENT AIR

Establishment Data
<u>Social Reason</u>: Faculty of Agricultural Sciences of the National University of the Northeast (UNNE).
<u>Address</u>: Juan Bautista Cabral N° 2131
<u>Location</u>: Corrientes
<u>Province</u>: Corrientes

ZIP CODE: 3400	<u>VAT ID NO</u>: 30 - 99900421 - 7

<u>Measurement Data</u>

<u>Make, model and serial number of the instrument used</u>:

Tenma's Partfold Counter, Model ST - 9880 and Serial No. 170107610

Date of the calibration certificate of the instrument used in the measurement: 16/05/2022

Date of measurement: 01/06/2022   Start time: 16:   00End time   : 18:00

<u>Atmospheric conditions</u>:

Relative humidity = 47,4 % Relative humidity = 47,4 % Relative humidity = 47,4

Temperature = 17,4 °C

Pressure = 1008,6 hPa

<u>Usual working hours/shifts</u>:

Wednesdays from 16:00 to 20:00

<u>Describe the normal and/or usual working conditions</u>:

Subject Industrial Hygiene and Safety in the Fifth Year of the Industrial Engineering Degree Course.

<u>Describe the working conditions at the time of measurement</u>:

Environmental measurements were carried out in the different sectors of the campus: access to the campus and the right side of the campus. These are used to measure the number and size of particles in the air. The data obtained is used to measure the contamination due to the fine dust present in the air. To calculate the data it draws in air for a selected time and calculates the number and size of particles contained in the air. In doing so, it considers equally the particle sizes 0.3, 0.5, 1.0, 2.5, 5.0 and 10.0 urn. The ambient contamination for a particle of the size selected by the user is indicated by a colour scale (see ISO Standard No. 14644 - 1: 2015 page 4); all measurements are within the green (permitted) colour range of this standard. In addition to the number of counted particles, the temperature, the air humidity and the dew point calculated on the basis of them, and the wet bulb temperature are indicated (see Figure N° 1 page N° 9).

<u>Documentation to be attached to the Measurement</u>

Drawing or sketch: Annex I (see Figures N° 2 and 3 page N° 8)

Figures: Annex I (see Figures N° 4 and 5 page N° 9)

2.1.1- Access to the campus (this measure was taken into account for the writing of the final paper).

<u>Environmental conditions</u>:

❖ Air Temperature = 18,6 °C (AT)

❖ Relative Humidity = 47,1 % (RH)

❖ Dew Point Temperature = 7,3 °C (DP)

❖ Wet Bulb Temperature = 13,2 °C (WB)

❖ Barometric Pressure = 1008,6 hPa

❖ Flow Rate = 2.83 litres/minute

<u>Measurement of particulate matter</u>:

❖ Channels (Size) = 0,3 - 0,5 - 1,0 - 2,5 - 5,0 - 10,0 um (ppm)

❖ p;irtfciil;is counting mode = Differential

<u>Results</u>:

Figure N° 1

- 0,3 um = 1315
- 0.5 um = 669
- 1.0 um = 125
- 2.5 |im = 12
- 5,0 um = 3
- 10.0 |im = 5

Complies with ISO Standard No. 14644 - 1: 2015 Green (Permitted).

<u>Plan or sketch: Annexes I, II and III:</u>

Figure N° 2

Figure N° 3

<u>Figures: Annex I:</u>

> Figures 4 and 5.

Figure N° 4 Figure N° 5

2.2- AMBIENT MULTIGAS MEASUREMENT

<u>Measurement Data</u>

<u>Make, model and serial number of the instrument used</u>:

Gas Detector, Drager Brand, Model X - am 2500 and Serial No. 8323918

Date of the calibration certificate of the instrument used in the measurement: 09/05/2022

Date of measurement: 01/06/2022   Start time: 16: 00End time   : 18:00

<u>Atmospheric conditions</u>:

Relative humidity = 47,4 % Relative humidity = 47,4 % Relative humidity = 47,4

Temperature = 17,4 °C

Pressure = 1008,6 hPa

<u>Usual working hours/shifts</u>:

Wednesdays from 16:00 to 20:00

<u>Describe the normal and/or usual working conditions</u>:

Subject Industrial Hygiene and Safety in the Fifth Year of the Industrial Engineering Degree Course.

<u>Describe the working conditions at the time of measurement</u>:

Environmental measurements were carried out in the different sectors of the campus, namely: access to the campus and the right side of the campus. It is used for continuous monitoring of the concentration of various gases in the ambient air, this concentration is measured in ppm (parts per million). Independent measurement of up to 6 gases corresponding to the Drager sensors installed. They are: Oxygen (O_2), Carbon Monoxide (CO), Hydrogen Sulphide (SH_2), Methane (CH_4), Nitrogen Dioxides (SO_2) and Sulphur Dioxides (SO_2) (see Annex III of Regulatory Decree N° 351/79 of Law N° 19587/72).

<u>Documentation to be attached to the Measurement</u>

Drawing or sketch: Annex II (see Figures N° 2 and 3 page N° 10)

Figures: Annex II (see Figures N° 6 and 7 page N° 14)

2.2.1- Access to the campus (this measurement was taken into account for the writing of the final paper) and the right side of the campus.

Results:

MULTIGAS MEASUREMENT								
[1] in ppm (parts per million) [2] LEL (Lfmite Lower Explosivity) [‡] Percent Volume O2 N/D (Not Detected Values)		ACCEPTABLE CONCENTRATIONS		MEASUREMENT LOG				
				1	2	3	4	5
PARAMETERS	Value Found	CMP	CMP - CPT	Hs: 17:00	Hs: -	Hs: -	Hs: -	Hs: -
Methane (CH_4) [ ]§	N/A	-	-	N/A	N/A	N/A	N/A	N/A
Oxygen (O_2) [***]	20,9	-	-	20,9	N/A	N/A	N/A	N/A
Carbon Monoxide (CO) [ ]**	N/A	25	-	N/A	N/A	N/A	N/A	N/A
Sulphur of Hydrogen (S_2H) [*]	N/A	10	15	N/A	N/A	N/A	N/A	N/A
Dioxide Nitrogen (NO_2) [*].	N/A	3	5	N/A	N/A	N/A	N/A	N/A
Dioxide Sulphur (SO_2) [*]	N/A	2	5	N/A	N/A	N/A	N/A	N/A

Normal concentration of oxygen (O_2) in the air.

"Normal oxygen level in the air: 20,9 % Volume "

Deficiency $\Longrightarrow$ Breathing.

Combustion, oxidation, inerting:

- Alarm level: 19,5 % Vol.

- Critical level: 16,0 % Vol.

"Enrichment $\Longrightarrow$ Fire.

Oxy-fuel cutting equipment:

- Alarm level: 23,5 % Volume

- CMP and CMP - CPT is applicable to toxic gases such as: Ammonia, Carbon

Monoxide, Chlorine, Hydrogen Cyanide, Hydrogen Sulphide, Nitric Oxide, Sulphur Dioxide, etc.

13

Accepted values of Annex III of the Regulatory Decree No. 351/79 of Law No. 19587/72.

<u>Figures: Annex II</u>:

> Figures 6 and 7.

Figure N° 6

Figure N° 7

<u>Measurement Data</u>

<u>Make, model and serial number of the instrument used</u>:

Mini Environmental Quality Meter, Brand Sper Scientific and Model No. 850027

Date of the calibration certificate of the instrument used in the measurement: 25/03/2022

Date of measurement: 01/06/2022   Start time: 16:00 End time: 18:00

<u>Atmospheric conditions</u>:

Relative humidity = 47,4 % Relative humidity = 47,4 % Relative humidity = 47,4

Temperature = 17,4 °C

Pressure = 1008,6 hPa

<u>Usual working hours/shifts</u>:

Wednesdays from 16:00 to 20:00

<u>Describe the normal and/or usual working conditions</u>:

Subject Industrial Hygiene and Safety in the Fifth Year of the Industrial Engineering Degree Course.

<u>Describe the working conditions at the time of measurement</u>:

Environmental measurements were taken in the different sectors of the campus: access to the campus and the right side of the campus. The factors measured are: Air Speed (metres/second), Air Flow (CMM), Fresh Wind (°C), Relative Humidity (%), Rotating Point Temperature (°C), Wet Bulb Temperature (°C), Heat Index (°C), Barometric Pressure (hPa) and Altitude (metres) (see Table No. 2).

Documentation to be attached to the Measurement

Drawing or sketch: Annex III (see Figures N° 2 and 3 page N° 10)

Figures: Annex III (see Figures N° 8 and 9 page N° 17)

Table N° 2: Heat Index from the US National Weather

Service (NWS)

Effects of Heat liidice (Shade Values)		
°C	°F	Notes
27 ~ 32	80 ~ 90	Caution: Fatigue is possible with prolonged exposure and activity. Continued activity may result in heat cramps.
32 ~ 41	90 ~ 105	Extreme caution: Heat cramps, heat exhaustion and heat exhaustion are possible. Continuing with the activity could lead to heat stroke.
41 ~ 54	105 ~130	Danger: Heat cramps, and heat exhaustion are likely; heat stroke is likely with continued activity.
+ 54	+ 30	Extreme danger: Heat stroke is imminent.

Exposure to full sun may increase heat index values by up to 8 °C (14 °F).

2.3.1- Access to the campus (this measure was taken into account for the writing of the final paper).

Results:

- Air Speed = 0,6 metres/second (An)
- Air Flow = 0.024 CMM (AirFL)
- Wind Chill = 17,2 °C (CHill)
- Relative Humidity = 50,1 % (RH)
- Rotio Point Temperature = 8,3 °C (dP)
- Wet Bulb Temperature = 12,1 °C (WET)
- Heat Index = 18,1 °C (HEAT) NORMAL
- Barometric Pressure = 1008,6 hPa (bAr)
- Altitude = 38 metres (HigH)

Figures: Annex III:

> Figures N° 8 and 9.

Figure N° 8 Figure N° 9

3. CONCLUSIONS.

Environmental measurements were carried out within the campus to determine the quantities and concentrations of particulate and multi-gas pollutants, as well as environmental quality, which can help to determine if there is an environmental problem (contamination). Locating the source of these various pollutants can help determine effective methods of reducing them and improving air quality. All measurements carried out comply with current international and national regulations.

4. REFERENCES.

ISO No. 14644 - 1 (2015), Standard.

NWS (National Weather Service), United States.

Occupational Health and Safety at Work (1972), Law No. 19587, Decree No. 19587, Decree No. 19587, Decree No. 19587, Decree No. 19587.

Regulatory (1979), No. 351, Annex III, Argentina.

SRT (Superintendencia de Riesgos del Trabajo), Web page, www.srt.gob.ar, Argentina.

Acknowledgements

o To my father Miguel and my mother Lida who gave me life and taught me the culture of study and work.

CHAPTER II

B-FINAL WORK EVALUATION OF AUDIBLE NOISE ACCORDING TO IRAM STANDARD NO. 4062/2016 (NOISE NUISANCE TO THE NEIGHBOURHOOD).

Ing. Qco. Jose Martm Glavas

Campus Facultad de Ciencias Agrarias de la Universidad Nacional del Nordeste (UNNE) - City of Corrientes - June 2023

josemartinglavas18@gmail.com

SUMMARY

Technical report of audible noise evaluation in the Campus of the Faculty of Agricultural Sciences of the National University of the Northeast (UNNE) carried out by the subject Industrial Hygiene and Safety of the Fifth Year of the Industrial Engineering Degree. In the measurements, the technical specifications of the IRAM N° 4062/2016 standard for the evaluation of noise nuisance to the neighbourhood and the sound measurement procedures were respected. These measurements were carried out at the access to the campus and in the corridor access to classroom A5, for noise monitoring a sound analyser (decibel meter), brand Sper Scientific and model No. 850069 was used.

Prior to the start of the tests, the correct functioning of the equipment used was verified by applying an acoustic calibrator. In each position, a series of measurements comprising a minimum number of readings was taken, in order to ensure that the results were representative. These readings corresponded to the equivalent continuous sound level (L_{Aeq}), evaluated over a period of 5 minutes, in compensated decibels A (dBA) and with the instrument positioned in response S (slow).

The decibel meter used is placed at a height of between 1.2 and 1.5 m above ground level and mounted on a tripod for these measurements. The students in this subject focused on processing the information collected in the field and determining the degree of adequacy of the noise levels generated in relation to the applicable regulatory requirements in force.

Keywords: Assessment, Noise, Audible, Environmental, Industrial.

1. INTRODUCTION.

Noise is one of the most common occupational and environmental pollutants. Large numbers of people are exposed daily to noise levels that are potentially hazardous to their hearing, in addition to suffering other detrimental effects on their health.

In many cases it is technically feasible to control excessive noise by applying acoustic engineering techniques to the sources of noise.

Among the effects suffered by people exposed to noise:

X Hearing loss.

X Tinnitus (ringing or buzzing).

X Interference in communication.

X Discomfort, stress, nervousness.

X Digestive system disorders.

X Cardiovascular effects.

X Decrease in work performance.

X Increase in accidents.

X Changes in social behaviour.

<u>Sound</u>:

Sound is a phenomenon of mechanical disturbance, which propagates in an elastic material medium (air, water, metal, wood, etc.) and which has the property of stimulating an auditory sensation.

<u>Noise</u>:

From a physical point of view, sound and noise are the same thing, but when the sound becomes unpleasant, when it is unwanted, it is called noise. In other words, the definition of noise is subjective.

<u>Health effects of noise</u>:

To understand the effects of noise it is useful to define the term health, according to the World Health Organisation. "Health is a state of complete physical, mental and social well-being and not merely the absence of disease or infirmity". The quotation is from the Preamble to the Constitution of the World Health Organization, which was adopted by the International Health Conference, held in New York from 19 June to 22 July 1946, signed on 22 July 1946 by the representatives of 61 States (Official Records of the World Health Organization, No. 2, p. 100), and entered into force on 7 April 1948.

It is also necessary to understand that noise is also associated with the environment and the activity carried out; it may seem obvious but the noise level tolerated in a dance hall has little or nothing to do with the noise level tolerated in a hospital or a school. This circumstance merits that when talking about noise and its effects, it is generally necessary to clarify the environment where the activity takes place and the duration of the activity.

For the purposes of this text it is necessary to define Environmental Noise in the terms of the WHO as: *"...the noise emitted by all sources except industrial areas. The main sources of urban noise are automobile, rail and air traffic, construction and public works and the neighbourhood. The main sources of indoor noise are ventilation systems, office machines, household appliances and neighbours. Characteristic neighbourhood noise comes from venues, such as restaurants, cafes, discotheques, etc.; live or recorded music; sports competitions (motor sports), playgrounds, car parks and pets, such as barking dogs. Many countries have regulated urban noise from aircraft and car traffic, construction machinery and industrial plants through emission standards and regulations for the acoustic properties of buildings. But few countries have regulations for neighbourhood urban noise, probably because of the lack of methods to define and measure it and the difficulty of controlling it. In large cities around the world, people are increasingly exposed to urban noise from the sources mentioned above and its health effects are seen as a growing problem".*

With regard to health effects, the World Health Organisation notes:

Speech Perception Interference. A large part of the population is susceptible to interference in speech communication and belongs to a vulnerable subgroup. The most sensitive are the elderly and the hearing impaired. Even mild hearing impairments in the high frequency band can cause problems with speech perception in a noisy environment. From the age of 40 onwards, people's ability to interpret difficult spoken messages with little linguistic redundancy deteriorates compared to people in their 20s and 30s.

It has also been shown that higher noise levels and greater reverberation have a greater effect on children (who have not yet completed language acquisition) than on young adults.

When listening to complicated messages (at school, in a foreign language or in a telephone conversation), the signal-to-noise ratio should be at least 15 dB at a speech level of 50 dBA. This noise level corresponds on average to a casual voice level for men and women located one metre apart. Consequently, for clear speech perception, the background noise level should not exceed 35 dBA. In classrooms or conference rooms, where speech perception is of great importance, or for sensitive groups, background noise levels should be as low as possible. Reverberation time of less than 1 second is also necessary for good speech communication in smaller rooms. For sensitive groups, such as the elderly, a reverberation time below 0.6 seconds is recommended for adequate speech communication, even in a quiet environment.

Hearing impairment. Hearing-impairing noise is not restricted to occupational situations. Outdoor concerts, discotheques, motorised and shooting sports, loudspeakers or recreational activities also have high noise levels. Other important sources are hearing aids, as well as toys and fireworks that emit impulse noise.

Sleep disorders. The measurable effects of noise on sleep start at LAeq of 30 dBA and above. However, the more intense the background noise, the greater the effect on sleep. Sensitive groups include mainly the elderly, shift workers, people with physical or mental disorders and other individuals with sleep difficulties.

The sleep disturbance due to intermittent noise events increases with the maximum noise level. Even if the total equivalent noise level is quite low, a few noise events with a high maximum sound pressure level will affect sleep. Therefore, to avoid sleep disturbance, urban noise standards should be expressed in terms of the equivalent noise sound level, the maximum noise levels and the number of noise events. It should be noted that low-frequency noise, e.g. from ventilation systems, can disturb rest and sleep even at low sound pressure levels.

When noise is continuous, the equivalent sound pressure level should not exceed 30 dBA indoors, if negative effects on sleep are to be avoided. Even for noise with a large proportion of low frequency sounds, a lower gma value is recommended. When background noise is low, noise above 45 dB LAmax should be limited and for sensitive people a much lower emite is preferred. Noise mitigation in the early part of the night is believed to be an effective means of helping people to fall asleep. It should be noted that the effect of noise depends in part on the nature of the source. A special case is newborns in incubators, for whom noise can cause sleep disturbances and other health effects.

Reading acquisition. Chronic noise exposure during early childhood can hinder reading acquisition and reduce motivation. Evidence indicates that the greater the exposure, the greater the damage. There is recent concern about concomitant physical and physiological changes (blood pressure and stress hormone levels). There is still insufficient information on these effects to establish specific guideline values.

However, it is clear that kindergartens and schools should not be close to significant sources of noise, such as roads, airports and factories.

Annoyance. The ability of a noise to cause annoyance depends on its physical characteristics, including sound pressure level, spectrum and variations of these properties over time. During the day, few people are highly disturbed by L_{Aeq} levels below 55 dBA, and few are moderately disturbed by L_{Aeq} levels below 50 dBA.

Sound levels during the afternoon and evening should be 5 to 10 dB lower than during the day. Noise with low frequency components requires lower guide values. For intermittent noise, the maximum sound pressure level and the number of noise events should be considered. Guidelines or measures to reduce noise should also take into account outdoor residential activities.

Social behaviour. The effects of environmental noise can be determined by assessing its interference with social behaviour and other activities. Urban noise that interferes with rest and recreation appears to be the most important. There is consistent evidence that noise above 80 dBA reduces cooperative attitudes and that loud noise also increases aggressive behaviour in individuals predisposed to aggression.

There is also concern that chronic high noise levels contribute to feelings of helplessness among school children. Further research is needed to develop guidelines on this issue and on cardiovascular and mental effects.

Regarding the consideration according to the place where the noise manifests itself, it can be identified as follows:

Housing. The effects of noise in housing are sleep disturbance, annoyance and interference with conversation. In bedrooms, the critical effect is sleep disturbance. The guideline values for bedrooms are 30 dB L_{Aeq} for continuous noise and 45 dB L_{Amax}

for single noise events. Lower noise levels can be annoying depending on the nature of the source. At night, outdoor sound levels at one metre from the facades of houses should not exceed 45 dB L_{Aeq} so that people can sleep with their windows open.

This value was obtained by assuming that the reduction of outside noise passing indoors through an open window is 15 dB. In order to converse without interference indoors during the day, the noise level should not exceed 35 dB L_{Aeq}. The maximum sound pressure level should be measured with the sound pressure meter set to "fast". In order to protect the majority of people from noise nuisance during the day, the outdoor sound level from continuous noise shall not exceed 55 dB L_{Aeq} on balconies, terraces and outdoor areas. During the day, the level of moderately annoying noise should not exceed 50 dB L_{Aeq}. Where practical and feasible, the lowest outdoor sound level should be considered as the maximum sound level advisable for a new event.

Schools and pre-schools. In schools, the critical effects of noise are interference with oral communication, disturbance of information analysis (e.g. reading comprehension and acquisition), communication of messages and annoyance. In order to be able to hear and understand spoken messages in the classroom, the background sound level should not be higher than 35 dB L_{Aeq} during lessons. For hearing-impaired children, an even lower sound level may be required. The reverberation time in the classroom should be 0.6 seconds and preferably shorter for hearing-impaired children.

In meeting rooms and school cafeterias, the reverberation time should be less than 1 second. In playgrounds, the sound level of noise from external sources should not exceed 55 dB L_{Aeq}, the same value given for outdoor residential areas during the day. For pre-schools the same critical effects and guideline values apply as for schools. During sleeping hours in sleeping rooms in pre-schools the guideline values for residential sleeping rooms shall be applied.

Hospitals. For the majority of hospital spaces, the critical effects are sleep disturbance, annoyance and interference with speech communication, including alarm signals. The L_{Amax} of sound events at night should not exceed 40 dBA indoors. For hospital wards, the indoor guideline value is 40 dB L_{Amax} at night. During the day and evening, the indoor guideline value is 30 dB L_{Aeq}. The maximum level should be measured with the sound pressure meter set to "fast".

Because patients are less able to cope with stress, the L_{Aeq} level should not be higher than 35 dB in most rooms where patients are treated and checked. Attention should be paid to sound levels in intensive care units and operating theatres. Incubators with sound inside can cause health problems for newborns, including sleep disturbance and hearing impairment. Further research is needed to establish sound levels in incubators.

Ceremonies, festivals and recreational events. In many countries, ceremonies, festivals and regular events are held to celebrate certain events. These events generally produce loud sounds, including music and impulse sounds. There is concern about the effect of loud music and impulse sounds on young people who frequently attend concerts, discos, video arcades, cinemas, amusement parks and outdoor events. At these events, the sound level generally exceeds 100 dB L_{Aeq}. Such exposure can lead to significant hearing impairment after frequent attendance.

In these premises, occupational exposure of employees should be regulated and at a minimum, the same standards should apply to customers. Customers should not be exposed to sound levels above 100 dB L_{Aeq} for a period of four hours more than four times a year. To avoid acute hearing impairment, the L_{Amax} should always be below 110 dB.

Hearing aids. To avoid hearing impairment caused by music through headphones in adults and children, the 24-hour equivalent sound level should not exceed 70 dBA. This implies that, for a daily exposure of one hour, the L_{Aeq} level should not exceed 85 dBA. To avoid acute hearing impairment, the L_{Amax} should always be below 110 dBA. Exposures are expressed with the equivalent sound level in the free field.

Toys, fireworks and firearms. To avoid acute mechanical damage to the inner ear caused by impulse sounds from toys, fireworks and firearms, adults should never be exposed to sound pressure levels of more than 140 dBA. For children at play, the maximum sound pressure level produced by toys should not exceed 120 dBA, measured close to the ear (100 mm). To avoid acute hearing impairment, the L_{Amax} should always be below 110 dBA.

Parks and conservation areas. Quiet outdoor areas should be preserved and a low signal to noise ratio should be maintained.

2. EQUATIONS AND TABLES.

The noise evaluation is carried out according to IRAM N° 4062/2016 (noise nuisance to the neighbourhood) from the measurement of the equivalent continuous sound level

(L_{Aeq}), response time S (slow: slow), for the reference times, affected by the correction factors:

$L_{Aeq} = 10Лоð((1/T).^(к.10^{u//o} \circ)$ *(Measured by decibel meter)*

Con:

L_i measured sound level, t_i measurement interval at constant noise, T total measurement period.

Correction for tonal and/or impulsive character: K_I

Corrected evaluation level: $L_E = L_{Aeq} + K_I$

Background noise level: L_F

Calculated sound level L_C: $L_C = L_b + K_z + K_u + K_h$

Con:

Base level (in decibels compensated A): $L_b = 4o\ dBA$

Correction factor for zone type: K_z

Correction factor for location in space to be assessed: K_u

Correction factor for timetable: K_h

$K_T + K_I + K_{BF}$ [dBA].	K [dBA] [dBA
0	0
5	5
7	6
10	6
12	7
15	Noise is BURDENING
17	Noise is BURDENING

Table N° 2 - Values of the correction calf, K_z

Zone	Type	T6correction term, K_z [dBA].
Hospital, rural residential	1	-5
Suburban with little traffic	2	0
Urban residential	3	5
Urban residential with some light industry or main roads*.	4	10
Intermediate commercial or industrial centre between types 4 and 6	5	15
Predominantly industrial, with few dwellings	6	20
* An urban residential area with light industry working only during the day will be type 3.		

Table N° 3 - Correction term values, K_u

Location on the farm	Correction term, K_u [dBA].
Indoors: premises bordering on the public highway	0
Premises not bordering the public highway	-5
Outdoors: uncovered areas not bordering the public highway. For example: gardens, terraces, patios, etc.	5

Table N° 4 - Values of tern one of correction, Kh

Period	Correction term, Kh dBA]
Working days: 08:00 to 20:00 hs Saturdays: 08:00 to 14:00 hs	5
Working days: 06:00 to 08:00 hs and 20:00 to 22:00 hs Saturdays: 14:00 to 22:00 hs Sundays and holidays: from 06:00 to 22:00 hs.	0
Night: from 22:00 to 06:00 hs	-5

Noise qualification:

$LE - LC < 8\ dBA$ *Not annoying noise*

$LE - LC > 8\ dBA$ *+ Nuisance noise*

Reference times:

The basis of the assessment is the characterisation of noise over three reference times whose start and end times, for the purposes of this regulation, are as follows:

Daytime hours Business days: 08:00 to 20:00 hs

Saturdays: 08:00 to 14:00 hs

Rest hours Weekdays: from 06:00 to 08:00 hs and from 06:00 to 08:00 hs.

20:00 to 22:00 hs

Saturdays: 14:00 to 22:00 hs

Sundays and public holidays: 06:00 to 22:00 hs

Night hours Night: 22:00 to 06:00 hs

3. DEVELOPMENT.

Aqrn has complied with the technical specifications of the IRAM standard N°.

4062/2016 for the evaluation of noise nuisance to the neighbourhood and sound measurement procedures. The sound sensor (decibel meter) used is placed on a tnpode at a height of between 1.2 and 1.5 m above ground level to carry out these measurements. The students in this subject focused on processing the information collected in the field and determining the degree of adequacy of the noise levels generated in relation to the applicable regulatory requirements (see page 17).

<h1 align="center">MEASUREMENT OF AMBIENT AUDIBLE NOISE</h1>

Establishment Data
<u>Social Reason</u>: Faculty of Agricultural Sciences of the National University of the Northeast (UNNE).
<u>Address</u>: Juan Bautista Cabral N° 2131
<u>Location</u>: Corrientes
<u>Province</u>: Corrientes

CP 3400	<u>VAT ID NO</u>: 30 - 99900421 - 7

<u>Measurement Data</u>

<u>Make, model and serial number of the instrument used</u>:

Sound analyser (decibel meter), Sper Scientific, Model No. 850069 and Serial No. AI13028. Complies with the following Standards: IEC 61872 - 1:2002 Class 2, IEC 61260:1995 Class 2, ANSI S1.11 - 2004 Class 2, ANSI S1.4 - 1983 Type 2.

Date of the calibration certificate of the instrument used in the measurement: 27/03/2023

Date of measurement: 07/06/2023   Start time: 17:   00End time   : 19:00

<u>Atmospheric conditions</u>:

Relative humidity = 79,0 % Relative humidity = 79,0 % Relative humidity = 79,0

Temperature = 24,0 °C

Pressure = 1007,6 hPa

<u>Usual working hours/shifts</u>:

Wednesdays from 16:00 to 20:00

<u>Describe the normal and/or usual working conditions</u>:

Subject Industrial Hygiene and Safety in the Fifth Year of the Industrial Engineering Degree Course.

<u>Describe the working conditions at the time of measurement</u>:

Environmental measurements were carried out in the different sectors of this campus: access to the campus and corridor access to classroom A5 (see page 17).

<u>Documentation to be attached to the Measurement</u>

Plan or sketch: Annex I (see Figure N° 1 page N° 26)

Figures: Annex II (see Figures N° 2, 3 and 4 page N° 27)

3.1- Results.

- $K_I = 0$ if no tonal and/or impulsive characteristics are present.

- $K_z = 5$ dBA (zone: urban residential, type 3; see Table N° 2 p. 26)

- $K_u = 5$ dBA (outdoor areas; see Table N° 3 page 27)

- $K_h = 5$ dBA for daytime hours (weekdays: 08:00 to 20:00 hs - Saturdays: 08:00 to 14:00 hs; see Table N° 4 page 27).

- $L_{Aeq1} = 61.8$ dBA (campus access)

- $L_{Aeq2} = 58.7$ dBA (corridor access to classroom A5)

Daytime audible noise evaluation:

$L_{E1} = L_{Aeq1} + KI = 61,8 + 0 = 61,8$ dBA

$L_{E2} = L_{Aeq2} + KI = 58,7 + 0 = 58,7$ dBA

$L_C = L_b + K_z + K_u + K_h = 40 + 5 + 5 + 5 + 5 = 55$ dBA

- $L_{E1} - L_C = 61,8 - 55 = 6,8$ dBA < 8 dBARUIDONOMOLESTE

- $L_{E2} - L_C = 58,7 - 55 = 3,7$ dBA < 8 dBARUIDONOMOLEST

<u>Annex I:</u>

Figure 1

<u>Annex II:</u>

ý Figures 2, 3 and 4.

Figure N° 2 Figure N° 3

Figure 4

4. CONCLUSIONS.

Environmental measurements were carried out within this campus (access to the campus and corridor access to classroom A5) to determine the equivalent continuous sound level quantities (daytime audible noise) measured by the decibel meter and that can help determine if there is an environmental problem (acoustic contamination) within the faculty. According to the results obtained in both points, the values are less than 8 dBA considered not annoying noise to the neighbourhood; therefore, it complies with the standard IRAM N° 4062/2016.

5. REFERENCES.

Instituto Argentino de Normalization y Certification (IRAM) N° 4062/2016, Norma.

Occupational Health and Safety (1972), Law No. 19587, Regulatory Decree (1979), No. 351, Argentina.

SRT (Superintendencia de Riesgos del Trabajo), Web page, www.srt.gob.ar, Argentina.

Operating Manual, Environmental Quality Meter with Sound, Sper Scientific, Model No. 850069, Serial No. AI13028, USA.

Acknowledgements

o To my father Miguel and my mother Lida who gave me life and taught me the culture of study and work.

CHAPTER III

C- FINAL WORK EVALUATION OF THERMAL LOAD ACCORDING TO THE LAW OF HYGIENE AND SAFETY AT WORK N° 19587/72 AND ITS REGULATORY DECREE N° 351/79.

Ing. Qco. Jose Martm Glavas

Campus Facultad de Ciencias Agrarias de la Universidad Nacional del Nordeste (UNNE) - City of Corrientes - June 2024

josemartinglavas18@gmail.com

SUMMARY

Technical report of thermal load evaluation in the Campus of the Faculty of Agrarian Sciences of the National University of the Northeast (UNNE) carried out by the subject Industrial Hygiene and Safety of the Fifth Year of the Industrial Engineering Career. In the measurements, the technical specifications of Annex II, Article N° 60, Chapter N° 8 Thermal Load of the Law N° 19587/1972 of Hygiene and Safety at Work and its Regulatory Decree N° 351/1979 of the Argentine Republic have been respected. Since exposure to heat is capable of generating pathological reactions in the human body, an evaluation of the exposure to thermal load must be carried out; the assessment of both thermal stress and thermal tension can be used to evaluate the risk to the health and safety of people. These measurements were carried out in the Dean's Office (closed environment) and in the Access to the Projection Room (open environment), using a thermal load monitor, Tenmars, model No. TM - 1880.

To determine the location (height) of the equipment and number of readings, check the homogeneity of the temperature in the surroundings of the workplace at different heights (from floor level), preferably taking three readings simultaneously using irfpode and extensions:

a) Reading 1: 1.7 metres (Access to the projection room in standing position).

b) Reading 2: 1.1 metres (Decanato Seated Position).

The students of this subject focused on the processing of the information collected in the field, in relation to the applicable regulatory requirements in force.

Keywords: Assessment, Load, Load, Thermal, Health, Safety, Security.

1- INTRODUCTION.

ANNEX II - Article N° 60 - Chapter N° 8 Thermal Load - Decree N° 351/1979

1.1-Instruments to be used.

The apparatus listed below constitutes a minimum set for the evaluation of the thermal load, without excluding others which can efficiently fulfil the same objectives, provided that their results are verifiable with those obtained with the methodology laid down in these Regulations.

1.1.1- Globothermometer.

It consists of a hollow copper sphere, painted matt black, with a thermometer or thermocouple inserted in it, so that the sensitive element is located in the centre of the sphere, with a wall thickness of 0,6 mm and a diameter of approximately 150 mm. The reading shall be checked every 5 minutes, reading its graduation after the first 20 minutes until a constant reading is obtained.

1.1.2- Natural wet bulb thermometer.

The natural wet bulb temperature shall be measured by means of a thermometer, the bulb of which is covered with a cotton cloth. The thermometer shall be soaked in distilled water for not less than half an hour before the reading is taken, shall extend to approximately the length of the bulb and shall be immersed in a vessel containing distilled water.

1.2-Estimation of metabolic heat.

This is done by means of tables according to job position and degree of activity.

The metabolic heat (M) will be considered as the sum of the basal metabolism (MB), and the additions derived from the position (MI) and the type of work (MII), therefore:

$$M = MB + MI + MII$$

Where:

1.2.1- Basal metabolism (BM).

To be considered at MB = 70 W

1.2.2- Position-derived addition (MI).

Position of the body	MI (W)
Lying or sitting	21
Standing	42
Walking	140
Climbing the slope	210

1.2.3- Addition derived from the type of work.

Type of work	MII (W)
Light manual work	28
Heavy manual labour	63
Working with one arm: Light	70
Working with one arm: Heavy	126
I work with both arms: Light	105
Work with both arms: Heavy	175
Work with the body: Light	210
Work with the body: Moderate	350
Work with the body: Heavy	490
Work with the body: Very heavy	630

PERMISSIBLE LIMITS FOR THERMAL LOAD

Values given in °C - TGBH

Work and rest regime	Type of work		
	Lightweight (less than 230 W)	Moderate (230 - 400 W)	Heavy (over 400 W)
Continuous work	30,0	26,7	25,0
75 % work and 25 % rest, every hour	30,6	28,0	25,9
50 % work and 50 % rest, every hour	31,4	29,4	27,9
25 % work and 75 % rest, every hour	32,2	31,1	30,0

Continuous work: Eight hours a day (48 hours a week).

If the resting place determines a temperature below 24 °C (TGBH) the resting time may be reduced by 25 %.

Conversion: Kcal/h = 1,163 Watt

1.3-Thermal load assessment.

For the purpose of assessing the exposure of workers subjected to thermal stress, the

Globe Bulb Humid Bulb Temperature Index (GBHT) shall be calculated.

This calculation will be based on the following equations:

a- For indoor or outdoor locations without solar load TGBH = 0,7 TBH + 0,3 TG.

b- For outdoor locations with solar load TGBH = 0.7 TBH + 0.2 TG + 0.1 TBS.

Where:

TGBH: Globe Globe Bulb Humid Bulb Temperature Index.

TBH: Natural Wet Bulb Temperature.

TBS: Dry bulb temperature.

TG: Globe temperature.

Situations not covered by this regulation shall be resolved by the competent authority according to the best information available.

2- DEVELOPMENT.

THERMAL LOAD ASSESSMENT TECHNICAL REPORT

Establishment Data
Social Reason: Faculty of Agricultural Sciences of the National University of the Northeast (UNNE).
Address: Juan Bautista Cabral N° 2131
Location: Corrientes
Province: Corrientes

CP 3400	VAT ID NO: 30 - 99900421 - 7

Measurement Data		
Make, model and serial number of the instrument used:		
Thermal Load Monitor, Brand Tenmars, Model TM - 1880 and Serif No.　e180700515		
Date of the calibration certificate of the instrument used in the measurement: 19/01/2024		
Date of measurement: 12/06/2023	Start time: 16:00	Finalization time: 18:00

Atmospheric conditions: Relative Humidity = 62,2 % Temperature = 26,4 °C Pressure = 1004,1 hPa

Usual working hours/shifts:

Wednesdays from 16:00 to 20:00

Describe the normal and/or usual working conditions:

Subject Industrial Hygiene and Safety in the Fifth Year of the Industrial Engineering Degree Course.

Describe the working conditions at the time of measurement:

Measurements were carried out in the different sectors of this campus: Access to the Projection Room and the Dean's Office Sitting Position (see page 43).

Documentation to be attached to the Measurement
Drawing or sketch: Annex I (see Figure N° 1 p. 44)
Figures: Annex II (see Figures N° 2, 3 and 4 page N° 45)

2.1-Sampling strategies.

The workstations sampled, the number and duration of measurements and the equipment used have been selected accordingly:

- Information provided by employees (teaching and non-teaching staff of the faculty).
- The description of tasks and exposure times provided by the institution.
- The technical criteria in accordance with the regulations in force.

2.2-Measurement specifications:

a) Reading 1: 1.7 metres ((Access to projection room in standing position).

b) Reading 2: 1.1 metres (Decanato Seated Position).

2.3-Estimation of metabolic heat per workstation.

2.3.1- Access to the Screening Room Position Parado.

<u>Environmental conditions</u>:

- ❖ Relative humidity = 60,5 % Relative humidity = 60,5 % Relative humidity = 60,5
- ❖ Air Temperature = 26,5 °C

$M = MB + MI + MII$

Where:

$MB = 70$ W, $MI = 42$ W and $MII = 105$ W

$M = (70 + 42 + 105)\ W = 217\ W$

2.3.2- Deanery Seated Position.

<u>Environmental conditions</u>:

- ❖ Relative Humidity = 63,8 % Relative Humidity = 63,8 % Relative Humidity = 63,8
- ❖ Air Temperature = 26,3 °C

$M = MB + MI + MII$

Where:

$MB = 70$ W, $MI = 21$ W and $MII = 105$ W

$M = (70 + 21 + 105)\ W = 196\ W$

Decree № 351/79 - ANNEX III - TABLE 2 - Selection criteria for exposure to thermal stress (TGBH values in °C)

Work requirements	Acclimatised			Not acclimatised			
	Slight	Moderate	Very heavy	Slight	Moderate	Heavy	Very heavy
100 % work	29,5	27,5	-	27,5	25	22,5	-
75 % work 25 % rest	30,5	28,5		29	26,5	24,5	
50 % work 50 % rest	31,5	29,5	27,5	30	28	26,5	25
25 % work 75 % rest	32,5	31	29,5	31	29	28	26,5

TABLE 3: EXAMPLES OF ACTIVITIES WITHIN

ENERGY EXPENDITURE CATEGORIES

Categories	Examples of activities
Reposada	- Sitting quietly. - Sitting with moderate arm movements.
Slight	- Sitting with moderate arm and leg movements. - Standing, with light to moderate work at a machine or table using mainly arms. - Using a table saw. - Standing, with light to moderate work on a machine or bench and some movement around it.

<table>
<tr><td colspan="5">THERMAL LOAD ASSESSMENT TECHNICAL REPORT</td></tr>
<tr><td colspan="4"><u>Social Reason</u>: Faculty of Agricultural Sciences of the National University of the Northeast (UNNE).</td><td><u>TAX IDENTIFICATION NUMBER</u> №: 30 - 99900421 - 7</td></tr>
<tr><td><u>Address</u>: Juan Bautista Cabral № 2131</td><td><u>Location</u>: Corrientes</td><td>ZIP CODE: 3400</td><td colspan="2"><u>Province</u>: Corrientes</td></tr>
</table>

Measurement Point	Sector	Temperature Dry Bulb (TBS) [°C].	Globe Temperature (TG) [°C] [°C	Wet Bulb Temperature (TBH) [°C].	Globe Bulb Wet Bulb Temperature (TGBH) [°C] Indoor/Outdoor	Referendum Value TGBH segtin Table 2	Category y Activity segtin Table 3	Complies (YES/NO)
Access to the Projection Room Position Parado	At 1.7 metres	26,5	26,4	21,0	22.7 (Outdoor)	30,5	Slight	YES
Posidon Seated Deanery	At 1.1 metres	26,3	26,2	21,4	22.9 (Inland)	30,5	Slight	YES

Figure N° 1

Figures: Annex II:

> Figures 2 and 3.

Figure N° 2 Figure N° 3

3- CONCLUSIONS.

3.1-At 1.7 metres (Access to the projection room Position Parado).

M = 217 W (metabolic heat)

Work and rest regime	Type of work
	Lightweight (less than 230 W)
Continuous work	30,0
75 % work and 25 % rest, every hour	30,6
50 % work and 50 % rest, every hour	31,4
25 % work and 75 % rest, every hour	32,2

TBGH (°C) found below the TBGH reference value (acclimatised and light) according to Table 2 and with categories (light) and activities (standing, light or moderate work at a machine or bench and some movement around it) according to Table 3.

3.2-At 1.1 metres (Decanato Seated Position).

M = 196 W (metabolic heat)

Work and rest regime	Type of work
	Lightweight (less than 230 W)
Continuous work	30,0
75 % work and 25 % rest, every hour	30,6
50 % work and 50 % rest, every hour	31,4
25 % work and 75 % rest, every hour	32,2

TBGH (°C) found below the TBGH reference value (acclimatised and light) according to Table 2 and with categories (light) and activities (sitting with moderate arm and leg movements) according to Table 3.

4- REFERENCES.

Occupational Health and Safety (1972), Law No. 19587, Regulatory Decree (1979),
No. 351, Argentina.

SRT (Superintendencia de Riesgos del Trabajo), Web page, www.srt.gob.ar,
Argentina.

Operation Manual, Thermal Load Monitor, Tenmars, Model TM - 1880 and Serial No.
180700515, Taiwan.

Acknowledgements

o   To my father Miguel and my mother Lida who gave me life and taught me the culture of study and work.

Buy your books fast and straightforward online - at one of world's fastest growing online book stores! Environmentally sound due to Print-on-Demand technologies.

Buy your books online at
www.morebooks.shop

Kaufen Sie Ihre Bücher schnell und unkompliziert online – auf einer der am schnellsten wachsenden Buchhandelsplattformen weltweit! Dank Print-On-Demand umwelt- und ressourcenschonend produziert.

Bücher schneller online kaufen
www.morebooks.shop

Printed by Books on Demand GmbH, Norderstedt / Germany